EXPLORING CAREERS WITHOUT COLLEGE

VOCATIONAL CAREERS IN CONSTRUCTION

by Cynthia Kennedy Henzel

BrightPoint Press

San Diego, CA

an imprint of ReferencePoint Press, Inc.
Printed in the United States

For more information, contact:
BrightPoint Press
PO Box 27779
San Diego, CA 92198
www.BrightPointPress.com

LIBRARY OF CONGRESS CATALOGING-IN-PUBLICATION DATA

Name: Henzel, Cynthia Kennedy, author.
Title: Vocational careers in construction / by Cynthia Kennedy Henzel.
Description: San Diego, CA: ReferencePoint Press, 2026 | Series: Exploring careers without college | Includes bibliographical references and index. | Audience: Grades 7–9
Identifiers: ISBN 9781678212704 (hardcover) | ISBN 9781678212711 (eBook)
The complete Library of Congress record is available at www.loc.gov.

CONTENTS

THE CONSTRUCTION INDUSTRY AT A GLANCE

SELF-EMPLOYMENT RATES IN THE CONSTRUCTION INDUSTRY, 2023

Less than 18% | 18% to 21% | 22% to 24% | 25% to 29% | 30% or higher

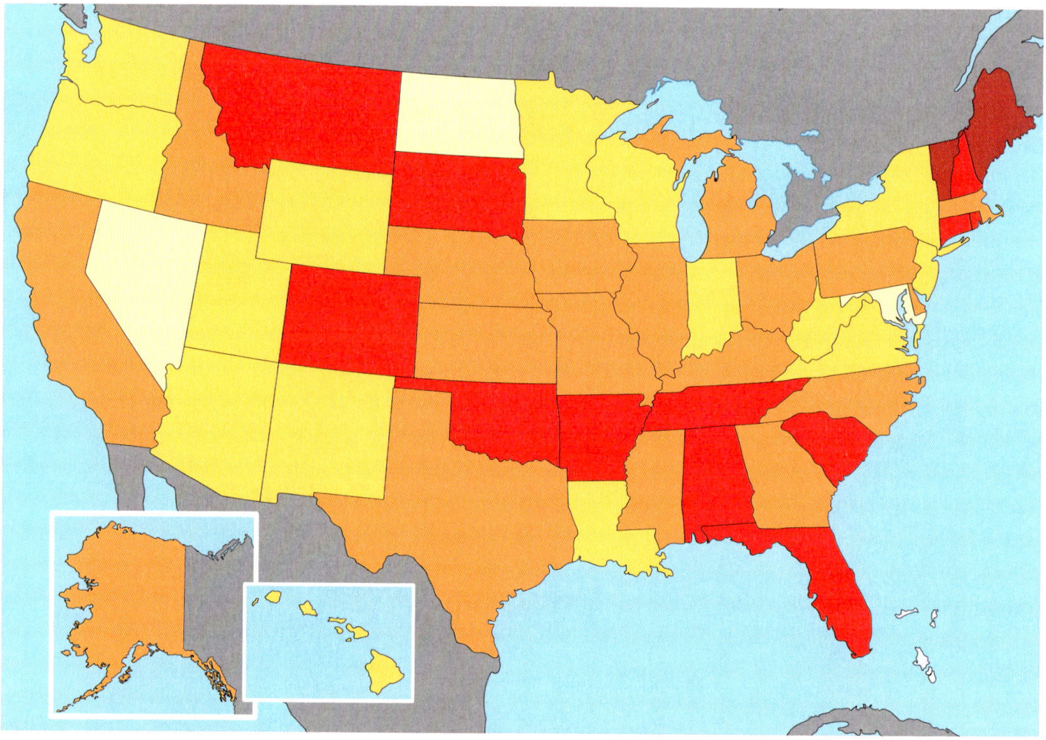

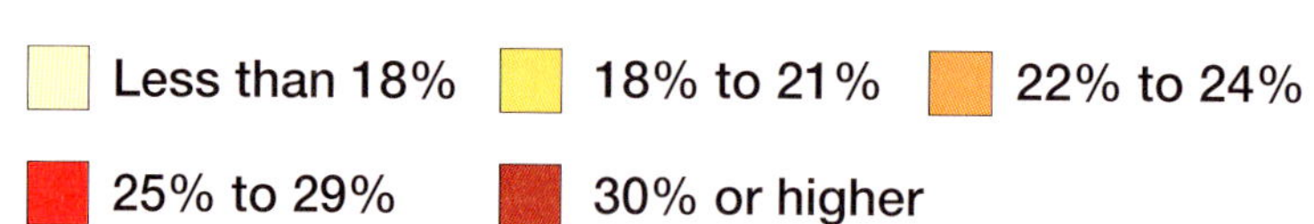

Source: Natalia Siniavskaia, "Construction Self-Employment Stable at 23%," National Association of Home Builders, *February 18, 2025, www.eyeonhousing.org.*

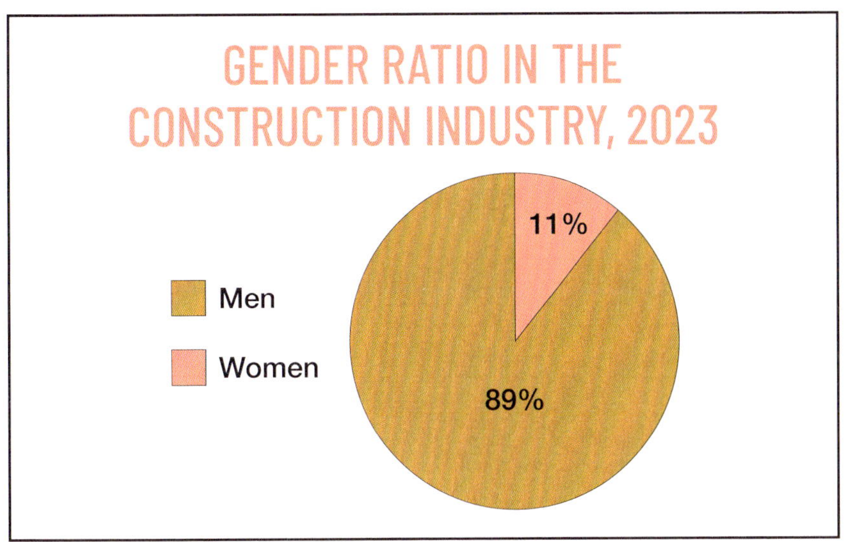

Source: "Gender Composition of the Construction Industry," National Center for Construction Education and Research, *2024, www.nccer.org.*

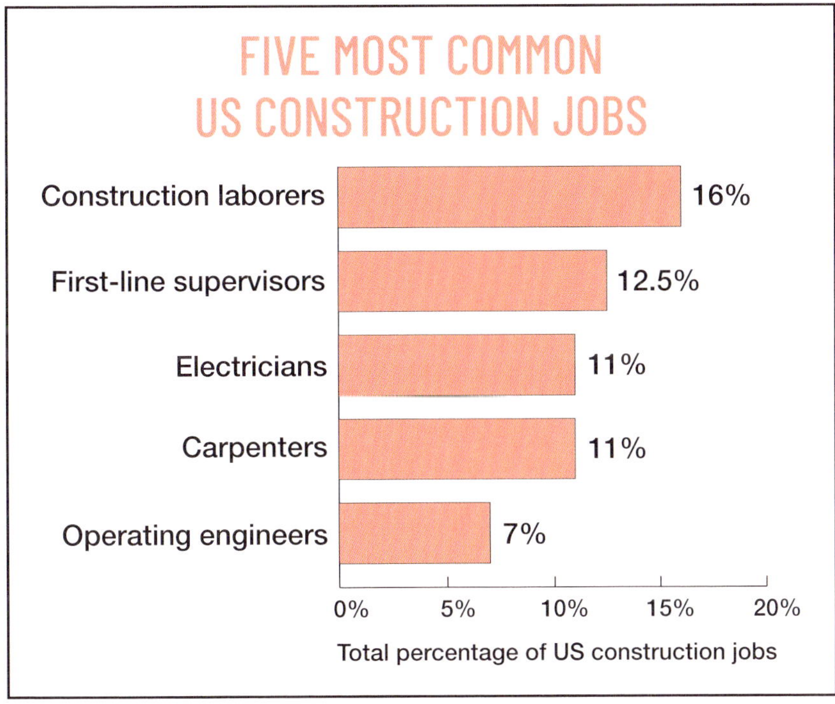

Source: Jonathan Jones, "The Most Popular Construction Jobs in the US," Construction Coverage, *February 5, 2025, www.constructioncoverage.com.*

WHAT IS THE CONSTRUCTION INDUSTRY?

Aaron Lussier stood on a narrow steel beam. Each day he balanced hundreds of feet in the air. It had taken him some time to get used to the height.

Lussier watched as a steel beam moved toward him. It was attached to a crane. The beam came closer. Lussier signaled the crane driver to guide it into place. He connected it to another beam. He checked

Construction workers often wear brightly colored jackets for safety reasons. Bright colors make it easier to see workers, even in bad weather and poor lighting.

On jobsites, workers use walkie-talkies to communicate with others. This ensures projects stay on time and helps prevent mistakes.

his safety harness. Then he walked to where the next piece would go.

Lussier was an ironworker. He was helping build a new bridge over the Anacostia River. The bridge went across a river in Washington, DC. Each day, the bridge got longer. Lussier found it rewarding to be a part of this work. In the end, the bridge was 1,600 feet (490 m) long. It took 4 years to complete it. Millions of people could now drive across the Frederick Douglass Memorial Bridge.

BUILDING THINGS

The construction industry is about building structures. Construction workers build with wood, steel, and concrete. Construction has many specialty jobs. Roofers work outside. So do concrete workers. Plumbing, electrical, and **drywall** are inside jobs.

Another type of construction is heavy and civil engineering. Jobs working with new technologies are fast growing as well. The Associated Builders and **Contractors** estimate that about 500,000 new construction workers are needed by 2026.

CHAPTER ONE

CARPENTER

Carpenters work with many types of materials. These include wood and plastic. They construct or repair buildings and furniture. Many build homes for people. Others work on larger projects. These might be office buildings or schools. The number of building jobs available often depends on how well the economy is doing.

Some carpenters specialize in the **framework** of buildings. They install rafters

Carpenters work with a variety of hand and power tools, such as hammers, drills, screwdrivers, and saws.

to support roofs and outside walls. Others finish the inside of a building. They put up interior walls. They install wood floors and put in cabinets.

Carpenters must pay attention to details. Accurate measurements are important.

Carpenter

Minimum Education: High school diploma
Personal Qualities: Physically strong, detail oriented, good communication skills
Certification and Licensing: Not required, but certifications can help advancement
Working Conditions: Carpentry is a physically demanding job, and workers may be inside or outside depending on the project.
Average Salary: $59,310 (2024)
Number of Jobs: 923,100 (2023)
Future Job Outlook: 4 percent growth expected between 2023 and 2033

Carpenters must accurately measure to make sure parts fit together correctly.

Math skills are necessary. Carpenters must calculate how much material is needed for a job. They use hand tools and power tools for different tasks. They must lift heavy materials. Carpenters need to be in good physical shape.

Strong communication skills are important, too. Carpenters communicate with coworkers and workers in different fields. This helps keep projects on schedule.

One part of a project must be finished before another part can begin. For example, carpenters cannot put in cabinets before the drywallers finish their job.

Carpentry students might learn alongside other students during training programs.

Carpenters must also talk with clients. They must discuss changes when something is not working. Sometimes, clients change their minds. Robert Kohn is a contractor. He says, “There are many types of carpentry and some require more accuracy, some more patience.”[1]

GETTING STARTED

Carpenters require a high school diploma. Math classes are important. Carpentry requires using fractions and simple math formulas. Some high schools offer a class in drafting. Drafting uses **precise** instruments to create building plans. These are known as blueprints. Carpenters must be able to read blueprints. The plans show how a structure is put together.

After high school, people usually get an apprenticeship. This is a training program. Apprentices work with an experienced carpenter. They begin using basic tools and doing simple tasks.

Apprentices earn between $33,000 and $46,000 in their first year. They can earn more as they gain experience. They may train with a private company. Others train under a journey-level carpenter. These carpenters have finished their apprenticeships and can work on their own.

The Carpenters Training Institute offers apprenticeships. Other organizations do, too. Students learn technical subjects in a classroom. They learn how to read blueprints. Students learn about safety measures and building codes as well.

Building codes are standards carpenters must follow. They keep structures safe. For example, carpenters must follow certain rules when building walls. Other rules apply when installing doors and windows. These rules make homes safer in events such as fires.

Apprenticeships generally last about 4 years. Most carpenters receive

Scenic Carpenter

Scenic carpenters build sets for movies or TV. They construct everything from spaceships to historic locations. Scenic carpenters use their skills to create buildings. They also create locations such as caves or laboratories. They work on riggings for lights, sound, and other equipment. Unlike most carpenters, scenic carpenters often tear down their work at the end of filming.

hands-on training. As apprentices learn, they do more work on their own. After 7,000 hours of training, an apprentice may become a journey-level carpenter.

Carpenters need an Occupational Safety and Health Administration certificate. They must take several exams to get it. Carpenters also need a driver's license to drive company vehicles. They might have to travel to different worksites as well.

ON THE JOB

Carpenters may work inside or outside. They spend many hours on their feet. They work on their knees to lay flooring. Some climb ladders to put up ceilings. They may also work in bad weather. Carpenters often go to work early. This may mean getting up

Carpenters should always take proper safety precautions, especially when carrying heavy equipment or working with dangerous tools.

before sunrise to get to a jobsite. They may work at one jobsite all day. Or they may go to several places.

Most carpenters work 8 hours a day. Sometimes, they work overtime. This might happen if a project falls behind. Carpenters who work outside may lose some workdays due to bad weather.

Carpenters take pride in what they create. Dan Champagne is a construction manager. He says, “I began my journey in carpentry because I have always had a passion for building things and seeing my creations come to life.”[2]

Carpenters have many opportunities to learn new skills. For example, some carpenters work in cabinetmaking. They use specific tools to make custom cabinets for homes or buildings.

Becoming a journey-level carpenter may be only the first step. Experienced carpenters often oversee a crew on a job. Some become responsible for a jobsite as a **foreman**. Others start their own business in remodeling or custom home building.

FIND OUT MORE

Carpenters Training Institute
www.carpenterstraininginstitute.org
The Carpenters Training Institute has a 4-year apprentice program. It includes classwork and on-the-job training. The institute also offers specialty apprenticeships in areas such as floor covering or pile driving. Pile driving is placing long columns of wood, steel, or concrete underground to support a structure.

Home Builders Institute
https://hbi.org
The Home Builders Institute is a nonprofit organization that trains students to work in the building industry. It provides training for pre-apprenticeship programs during high school. The institute also offers trade academies in carpentry at no cost to students.

CHAPTER TWO

TILE SETTER

Tile setters install tiles indoors and outdoors. They might also install stone. Tile can be made of ceramic, glass, marble, or granite. It is easy to clean. Tile is often used on floors. It is also used in areas that need protection from water. These include showers or kitchen walls. Tile is also used for swimming pools and fountains.

Tiles come in many colors and patterns. Setters learn different techniques to

Tile setters are also known as tile installers.

create beautiful tilework. One is scribing. This is measuring and cutting tile around an unusual surface or shape. It requires creativity and precision.

Tile Setter

Minimum Education: None, although a high school diploma is required for some apprenticeships
Personal Qualities: Good color vision, detail oriented, physical stamina
Certification and Licensing: Certifications are offered for years of experience and specialties.
Working Conditions: Being a tile setter involves hard physical labor, which includes spending a lot of time bending and kneeling.
Average Salary: $52,000 (2024)
Number of Jobs: 115,400 (2023)
Future Job Outlook: 6 percent growth expected between 2023 and 2033

Tile setters may work for large building contractors. Many work on their own. Melissa Swan's company focuses on bathrooms. She explains, "Showers are our bread and butter. . . . Everyone needs a shower. Turns out, people with a little more expendable income tend to like pretty showers. These are our clients."[3]

GETTING STARTED

Becoming a tile setter does not require formal education. But having a high school diploma is helpful for apprenticeships. Math is an important skill. Tile setters measure and work with fractions. They must also calculate how much tile is needed for a project. Art classes are useful for learning to create designs. Tile setters work with

Tile setters often have to place tiles individually. However, some small tiles may come in sheets to make the installation quicker.

different colors and patterns. They need to know which colors go well together. They must also follow designs.

Most tile setters learn on the job. They work with experienced tradespeople to learn basic skills. Some students may attend an apprenticeship trade school. They may go for 2 to 4 years.

Students learn about building codes and safety. They also learn how to read blueprints.

The Ceramic Tile Education Foundation offers a Certified Tile Installer certificate. To earn this certificate, workers must have 2 years of work experience. They must pass a written test. Then they must pass a hands-on test. This shows they have the skills for the job.

Mosaics

Mosaic tiles are small tiles that are put together to create designs or pictures. They are often used on walls. Some are placed at the bottom of swimming pools. Communities may put tile mosaics along highways. These artworks might tell a story about a community.

ON THE JOB

Tile setting is a multistep process. One tile installer said, "Most people have no idea about the thought, the labor, the planning, that goes into high-end tile work."[4] Each job means at least 1 hour of setup time.

The floors must be cleaned and leveled on the first day. If the job is a remodel, old floors are removed. Tile workers use a grinder to level high spots. They spread liquid leveler across low spots. The leveler takes a day to dry.

Planning is essential. Complicated patterns take time to lay out. Even using straight tiles requires planning. The tile setter calculates where tiles will end in different areas. They often lay out some rows first.

This ensures the spacing between the tiles works for a particular area.

Next, setters cut the tile with a wet saw. The saw has a circular blade. It runs through water to help cut hard materials.

Some workers use manual tile cutters, which work well for cutting ceramic and porcelain tiles that are flat and smooth.

Tile setters must follow designs so pieces are installed in the correct patterns.

Tile setters wear goggles. This protects their eyes from water and tile pieces. They make straight cuts or custom cuts to fit around corners. Handheld tile-cutting tools are used for small pieces.

Workers put on a dust mask after cutting the tiles. They mix a cement-like substance called thinset. This is done with a drill in a big bucket. The tile setter spreads the thinset evenly across the floor. Then they use a tool called a trowel to create ridges. These ridges make leveling tiles easier.

Setters press each tile into place. A small space is left between them. Setters must work quickly before the thinset hardens. After the tiles are placed, the thinset must completely dry.

The next day, tile setters can apply grout. Grout is a special material used to fill spaces. Setters use a rubber trowel to push grout between tiles. They then use a damp sponge to smooth the lines and remove extra grout. When the grout dries, they apply a layer of clear liquid sealer. This protects the grout and tiles.

Making a tiled surface takes a lot of work. But it can last many years. The finished product is often the highlight of a room or wall. Joshua Nordstrom is a tile setter. He says:

> *There are many benefits to being a tile craftsperson. For me, number one is a sense of pride in my work. It's truly gratifying to see the look on my clients' faces when a job is completed.*[5]

FIND OUT MORE

Advanced Certifications for Tile Installers
www.tilecertifications.com
Advanced Certifications for Tile Installers (ACT) offers extra certifications in areas such as large tile, shower, or grout. ACT is not a training program. Instead, it certifies that a tile setter has knowledge in a certain area.

Ceramic Tile Education Foundation
www.ceramictilefoundation.org
The Ceramic Tile Education Foundation provides training for tile setters of all skill levels. It also offers two tile installer certifications.

TileLetter
www.tileletter.com
TileLetter is an industry magazine. It contains news about new products, special projects, upcoming trainings, and everything of interest to tile setters.

CHAPTER THREE

PLUMBER

Plumbers install and repair water and gas pipe systems. They work with copper, steel, or special plastic pipes. They also install and repair water fixtures. These include sinks, bathtubs, and toilets. Plumbers must follow strict building codes and regulations.

For new installations, plumbers need to read blueprints. These show where pipes and fixtures are located. Plumbers often

Turning off water and gas lines during emergencies or repairs is an important part of a plumber's job.

work for building contractors. For repairs, plumbers might work directly with homeowners. Before they begin a repair, plumbers look at the problem. Then they provide an estimate of what the repair

Plumber

Minimum Education: High school diploma and vocational training or apprenticeship
Personal Qualities: Communication skills, physical strength, problem-solving skills
Certification and Licensing: Most states require a license.
Working Conditions: Plumbers often work in homes and businesses with pipes and related systems. They are sometimes on call for jobs.
Average Salary: $62,970 (2024)
Number of Jobs: 473,400 (2023)
Future Job Outlook: 6 percent growth expected between 2023 and 2033

will cost. This includes the cost of parts. It also includes payment for their time.

Plumbers need good problem-solving skills. Plumbing pipes are generally hidden inside walls or beneath floors. This means walls or floors may have to be cut open to work on the pipes. Sometimes pipes need to be rerouted.

Plumbing problems can be stressful for homeowners. Plumbers need to have good people skills. They need to know how to talk to customers about different issues. Plumber David Ellingsen says:

> *The work is actually satisfying. There's nothing like feeling you have spent the day doing something useful. Something that has a direct positive effect on people's day to day lives.*[6]

Plumbers should clearly communicate with clients and explain complex issues in a way that people can understand.

Some specialized plumbers are pipe fitters. They install pipes that carry chemicals or gases. For example, they might install pipes for heating and cooling systems. Pipe fitters mainly work for manufacturers or industries.

Steamfitters install pipes for industrial plants. They work with systems that use high-pressure liquids and gases. Steamfitters also install heat and air systems in businesses.

GETTING STARTED

Plumbers need a high school diploma. Math, physics, and computer classes are useful for this career. Some students may go to work as a plumber's helper after

Sprinkler Fitter

Sprinkler fitters install and repair systems that assist in case of a fire. These plumbing specialists put different types of systems in homes. They also install them in large buildings such as schools. Sprinkler fitters inspect and test existing systems. They make sure the systems are ready for emergency situations.

high school. They can help repair, maintain, and install plumbing systems. But before becoming a plumber, people need to do an apprenticeship.

Apprentice programs are available through **unions**. Plumbing trade associations and businesses also offer programs. Apprentices need 2,000 hours of on-the-job training. They must learn technical skills as well. Apprentices learn safety practices and local building codes. They learn how to read blueprints. An apprenticeship takes about 4 to 5 years.

Apprentices take an exam to become a journey-level plumber. They can then work independently. With experience, journey-level plumbers can take a second exam to become a master plumber.

Plumbing apprentices are usually paid. The amount of pay and whether it is hourly or salary depends on the state the apprenticeship is in.

This means higher pay. Some states require plumbers who run their own business to be master plumbers.

Fixing water leaks, unclogging pipes, and addressing low water pressure in appliances are just a few of the common repairs that plumbers do.

ON THE JOB

Plumbers do a variety of tasks. One morning it could be a blocked drain. The next they might be replacing a dishwasher. Every day is a new place and a new problem to solve. Part of the job is using special tools to locate problems.

These tools include drain cameras or heat-sensing instruments. Plumbers estimate the cost of repair. They also explain fixture options to the client. Then plumbers must gather parts for their jobs.

Plumbers must be in good physical condition. They must move heavy equipment such as drain snakes. These are heavy cables used to clean drains. Plumbers climb ladders to reach high places. They work in tight spaces under cabinets. They might also work outside during bad weather.

Injuries can happen when working with sharp tools. Burns from hot pipes or soldering are common. Soldering is using heat to join metal pieces together. Plumbers may work long hours to get a job done.

They are often on call for emergencies at night or on the weekends.

Plumbers fix anything concerning water. So they may be called to fix a variety of items from fountains to drains. Mike Boeddeker is a master plumber. He says, "I like being in a profession that takes care of people. . . . A lot of training and education goes into being a plumber."[7]

FIND OUT MORE

Explore the Trades: How to Become a Plumber

https://explorethetrades.org/what-we-do/education/plumbing/how-to-become-a-plumber

This page on the Explore the Trades website provides step-by-step instructions on becoming a plumber and moving up in the trade.

Plumbing-Heating-Cooling Contractors Association

www.phccweb.org

The Plumbing-Heating-Cooling Contractors Association (PHCC) offers many educational opportunities for people at all levels, from pre-apprentice students to experienced plumbers. The PHCC site also helps workers build connections with others in the industry.

CHAPTER FOUR

ELECTRICIAN

Electricians install and repair systems that require electricity. They work on wiring, lights, appliances, and other equipment. They read blueprints to install new wiring. Or they use blueprints to locate existing systems. Electricians use testing devices to detect electrical flow in wires. This helps find problems in wiring and motors.

Electricians work with a variety of hand tools, including pliers, screwdrivers, wire strippers, and cable cutters.

Electricians need good color vision. This is because electrical wiring is color coded. Putting the wrong wires together can be dangerous. Electricians need strong communication skills. Explaining electrical

Electrician

Minimum Education: High school diploma and apprenticeship

Personal Qualities: Good color vision, critical-thinking skills, good hand-eye coordination

Certification and Licensing: Most states require a license.

Working Conditions: Being an electrician involves hard physical work. Some have to work at great heights or with high-voltage systems.

Average Salary: $62,350 (2024)

Number of Jobs: 779,800 (2023)

Future Job Outlook: 11 percent growth expected between 2023 and 2033

systems to a homeowner can be difficult. So electricians need to know how to clearly share information. Electricians also need physical strength. Rolls of electrical wire are often heavy.

Electricians may work outside or inside. While outside, they might work on high electrical lines. These carry electricity long distances. Electricians are responsible for bringing back the electricity after power outages. Some work on renewable energy projects such as wind and solar power. Others work inside in homes or commercial buildings. Their jobs often require working in small spaces or on ladders.

Electricians are at risk of serious injuries. These include falls and burns. Since electricity is dangerous, electricians wear

Depending on the project, electricians may work by themselves or in groups.

special clothing. An electrical shock can be deadly. So they wear clothing that does not **conduct** electricity well. This includes rubber gloves and special inserts in their boots. Hard hats do not conduct electricity either. And safety goggles protect a worker's eyes from sparks.

When working on powerful electric lines, electricians wear an arc flash hood. It looks like a space suit. It protects a worker from heat and bright light caused by sparking wires. Electricians also use special hand tools that do not conduct electricity.

GETTING STARTED

Becoming an electrician requires a high school diploma. Students need math skills to figure out wiring lengths. They also need to calculate the strength of electrical currents. Basic physics classes are important as well. They help students understand the concept of electricity.

Some electrician students choose to attend a trade school. They must then do an apprenticeship. Apprentice programs require 8,000 to 10,000 hours of on-the-job training. This generally takes 4 to 5 years. Apprentices work with master electricians. They learn about the tools for testing electrical systems. They also install and repair electrical wiring.

Apprentices need 576 to 1,000 hours of technical training. This classroom training teaches electrical theory. Students learn to understand wiring diagrams. They read blueprints. Apprentices take classes in code requirements. They learn safety, too.

Trade organizations offer apprenticeships as well. One is the Independent Electrical Contractors. People interested in joining a union can contact the International Brotherhood of Electrical Workers. It also places workers into apprenticeships.

After their apprenticeship, electricians are required to pass a test to get a license. Each state has different tests. This is because states have their own electrical codes and regulations. Some require

Space Electrician

Space electricians work on satellites and spacecraft. They also work on structures at facilities for the National Aeronautics and Space Administration. These specialists need experience as licensed electricians. They can get extra training to learn about issues specifically related to space. These might be complex wiring or parts for planes and spacecraft.

electricians to continue taking education courses to keep their license.

Electricians move from site to site for work. They may drive company vehicles. This means they must have a driver's license.

A multimeter is a device electricians use to measure different electrical elements, such as current, resistance, and voltage.

ON THE JOB

Electricians often start their day early. They first check with a foreman to get a list of tasks for the day. They may need to replace wiring in a home. Or they may need to find the reason for flickering lights in a business. Putting in new appliances or fixing old ones is also part of an electrician's job. In emergencies such as power outages, electricians may work long hours. Keeping power on is an important service.

Electricians often have scheduled appointments. Staying on time with projects can be difficult to do. Finding and repairing wiring may be quick. Or it can take a lot of time. A wiring issue could affect many places in a house. Electricians have to check all those areas. Not rushing jobs is important. Electrician Phil Remster says, "Safety is the keystone of the electrician's profession. Each task must be approached with a **meticulous** regard for safeguarding oneself and others from electrical hazards."[8]

Journey-level electricians might supervise apprentices. After 3 years in journey-level positions, electricians can take a test to become master electricians. They may also go into specialties. These include solar or electrical generation.

Electricians are in high demand. They earn good salaries and have stable jobs. Electrician Michael Lucas enjoys the challenge of solving problems. He says, "Solving electrical problems can be like playing detective."[9]

FIND OUT MORE

Explore the Trades: How to Become an Electrician
https://explorethetrades.org/what-we-do/education/electrician/how-to-become-an-electrician
This page on the Explore the Trades website provides information on how people can become electricians. The site includes step-by-step guides and answers questions about the industry.

National Electrical Contractors Association
www.necanet.org
The National Electrical Contractors Association (NECA) is an electrical organization that provides services to buildings and housing across the United States. The NECA site has news about the industry and links to chapter associations in different states.

OTHER JOBS IN THE CONSTRUCTION INDUSTRY

HVAC Technician

Heating, ventilation, and air conditioning (HVAC) technicians work on heating and cooling systems. They inspect systems for problems. They also install new systems. Part of their job is to explain problems to customers. HVAC technicians often make suggestions to help systems work better. Or they may have a contract to regularly maintain a system. Most HVAC techs attend a trade school. This takes between 6 months and 2 years. HVAC techs can work with a professional in the field for hands-on training.

Wind Turbine Technician

Wind turbine technicians maintain and repair wind turbines. These convert wind into electricity. Technicians climb turbines to inspect and repair the towers and blades. They test the different parts of the turbine that generate power. Then they repair or replace parts as needed. In addition, technicians collect data for research. Wind turbine technician is one of the fastest-growing occupations. Demand is expected to increase by 60 percent between 2023 and 2033.

Ironworker

Ironworkers build with iron and steel. They construct large buildings and bridges. They also make other structures, such as stadiums. Some ironworkers cut and shape steel pieces. Others work high above the ground. Cranes lift the heavy beams up to the worker. The ironworker then welds or bolts the beam in place. Ironworkers have a dangerous job. But they use safety equipment to prevent accidents. The job takes physical strength. Most ironworkers learn their trade through apprenticeships.

Heavy Equipment Operator

Heavy equipment operators use large machines. They use bulldozers and backhoes to dig and move soil. They work with large machines to spread and flatten materials. And they operate forklifts and cranes. These machines lift heavy loads from one place to another. Heavy equipment operators maintain their machines and inspect them for safety problems. A license to operate heavy equipment is required. This can be acquired through technical school or apprenticeship.

GLOSSARY

conduct

to transfer heat or electricity

contractors

people or companies hired to oversee construction projects

drywall

a type of board made from plaster used to create walls inside a building

foreman

a supervisor who oversees a group of workers to ensure work is done safely and efficiently

framework

the part of a building that supports a structure

meticulous

giving a lot of attention to details by being careful and precise

precise

accurate or exact

unions

associations formed by people with common interests, often to advance trades

SOURCE NOTES

CHAPTER ONE: CARPENTER

1. Robert Kohn, "What Does It Take to Become a Carpenter? Robert's Answer," *CareerVillage.org*, June 13, 2023. www.careervillage.org.

2. Dan Champagne, "What Does It Take to Become a Carpenter? Dan's Answer," *CareerVillage.org*, March 8, 2024. www.careervillage.org.

CHAPTER TWO: TILE SETTER

3. Quoted in Lesley Goddin, "Women in Tile 2022," *TileLetter*, March 1, 2022. www.tileletter.com.

4. Quoted in "A Good Tile Installer Is King . . . & Other Revelations You've Never Thought About," *Hamilton Tile*, October 29, 2017. www.hamiltontilega.com.

5. Joshua Nordstrom, "NTCA Tile Setter Craftsperson of the Year, Residential," *TileLetter*, May 1, 2024. www.tileletter.com.

CHAPTER THREE: PLUMBER

6. Quoted in Tammy Sofranic, "A Day in the Life of . . . a Plumber," *Skillsroad*, n.d. https://skillsroad.com.au.

7. Quoted in "An Eye-Opening Day in the Life of a Professional Plumber," *Bears Home Solutions*, n.d. https://bearshomesolutions.com.

CHAPTER FOUR: ELECTRICIAN

8. Phil Remster, "A Day in the Life of an Electrician—an Insider's Look at This Rewarding Profession," *SJ Technical Training*, March 28, 2024. https://sjelectricaltraining.com.

9. Quoted in Tonia Nifong, "A Day in the Life of an Electrician: Michael Lucas," *Berwick Electric Co.*, October 16, 2013. www.berwickelectric.com.

INDEX

IMAGE CREDITS

Cover: © Robert Kneschke/Shutterstock Images
4: Red Line Editorial
5: Red Line Editorial
7: © Randy Hergenrether/Shutterstock Images
8: © PeopleImages.com-Yuri A./Shutterstock Images
11: © Shakirov Albert/Shutterstock Images
13: © Jannissimo/Shutterstock Images
14: © Monkey Business Images/Shutterstock Images
19: © anatoliy_gleb/Shutterstock Images
23: © BearFotos/Shutterstock Images
26: © Treetouch/Shutterstock Images
29: © Sergey Kuznecov/Shutterstock Images
30: © P A/Shutterstock Images
35: © Monkey Business Images/Shutterstock Images
38: © Dragon Images/Shutterstock Images
41: © Daisy Daisy/Shutterstock Images
42: © Ivant Weng Wai/iStockphoto
47: © Phovoir/Shutterstock Images
50: © Cat Box/Shutterstock Images
54: © AtlasStudio/Shutterstock Images
58 (top): © Hryshchyshen Serhii/Shutterstock Images
58 (bottom): © BaLL LunLa/Shutterstock Images
59 (top): © Parilov/Shutterstock Images
59 (bottom): © M2020/Shutterstock Images

ABOUT THE AUTHOR

Cynthia Kennedy Henzel has degrees in education and geography. She has written more than a hundred books for young people. These include fiction and nonfiction on subjects such as geography, science, culture, and history. Henzel has roots in construction. Her father built their home by hand—most of it while they were living there. She enjoys traveling. But when at home, she can be found remodeling houses.